Chipset's Circuitry's Choreography

The motherboard's complex orchestration and hardware component interplay

Robin Juárez

Contents

1 Introduction to Motherboards

The motherboard, often overlooked, is the unsung hero of any computer system. It's the foundation upon which everything else is built, a complex circuit board that orchestrates the intricate dance of hardware components. Understanding its architecture is crucial for anyone serious about building or repairing computers.

Think of it as a city, a bustling metropolis with various districts each playing a vital role. The CPU, the city's central power plant, sits at the heart, connected to the motherboard via a socket specifically designed for its type and size. This socket isn't just a simple connection; it ensures precise electrical contact, enabling high-speed data transfer between the CPU and the rest of the system. The size and type of socket are crucial, as they dictate which CPUs are compatible with a specific motherboard. Choosing the right combination is essential for optimal performance. Different chipsets support different processors, so careful consideration of the motherboard's chipset is vital for compatibility.

RAM, the city's short-term memory, lives in designated slots, typically DIMM slots. These slots allow for easy installation and removal of RAM modules, enabling upgrades and modifications. The type of RAM – DDR3, DDR4, or DDR5 – and its speed dictate the system's overall memory performance. More RAM means more simultaneous tasks can be handled effectively, but it's vital to select RAM that's compatible with the motherboard's specifications. Installing incompatible RAM can lead to system instability or even failure.

Expansion slots, like avenues and highways, enable the connection of additional hardware. PCIe slots are the main arteries, allowing for high-bandwidth connections to graphics cards, sound cards, and other peripherals. Knowing the types of PCIe slots and their bandwidths is crucial for choosing and installing compatible expansion cards. One should also consider the number of available PCIe slots and their physical placement on the motherboard, as this can impact airflow and overall system design. The placement can affect the size and type of graphics cards that can be used effectively.

Storage devices, the city's archives, connect via SATA or NVMe interfaces, located on the motherboard. The choice between these interfaces greatly influences data access speeds; NVMe, generally faster, is ideal for operating systems and frequently accessed applications. SATA, being more widespread and often more affordable, serves as a good option for mass storage. Understanding the differences in speed, reliability, and cost is important for making informed decisions about storage upgrades.

Furthermore, the motherboard's southbridge, the city's local government, handles lower-speed peripherals and devices like USB ports, audio jacks, and Ethernet connections. This crucial component manages various inputs and outputs, ensuring everything connects and functions correctly. It coordinates the interactions between the CPU and other lower-speed peripherals, making it an indispensable part of the entire system's functionality. Understanding the chipset's capabilities, particularly the functionalities of the northbridge and southbridge, is essential for selecting a motherboard that meets your specific needs and can support your chosen components.

The BIOS or UEFI, the city's central control system, is a firmware embedded in a chip on the motherboard. It's the first piece of software that runs when you turn on your computer. It's responsible for initializing the hardware and loading the operating system. Working with the BIOS or UEFI, while often overlooked, allows for optimization of system performance, enabling modifications to boot order, CPU settings, and various other parameters. Accessing and navigating these settings require some level of understanding, but it's crucial for experienced users to unlock the full potential of their system.

In conclusion, the motherboard is a complex piece of engineering, a critical component deserving of deeper understanding. By grasping its architecture, functionality, and interaction with other components, you'll be better equipped to build, maintain, and troubleshoot your computer systems effectively. Each component—from the CPU socket to the expansion slots—plays a crucial role, and understanding their interdependencies is vital for optimal performance and system stability. This knowledge isn't merely theoretical; it's practical, equipping you to make informed decisions when selecting components and troubleshooting issues. A solid understanding of motherboards is a cornerstone to advanced computer hardware knowledge.

1.1 Board Anatomy

The motherboard, often overlooked but undeniably vital, resembles a complex city map. Its intricate network of pathways, meticulously designed, guides the flow of data, power, and instructions throughout your entire computer system. Understanding this layout is fundamental to appreciating its function. Think of it as the backbone, the central nervous system, everything converges here.

Unlike the simplified diagrams often seen, real-world motherboards are densely packed with components. Each section, each connector, has a specific purpose. You'll find rows of sockets for RAM modules, neatly arranged and color-coded for easy identification. These slots are not interchangeable; specific types of RAM are designed for specific sockets. Incorrect placement can lead to system failure.

Then there's the CPU socket, a crucial element at the heart of it all. This is where the central processing unit, the computer's brain, physically resides. Its design varies depending on the CPU type and motherboard manufacturer. The socket itself is a highly precise connection ensuring the CPU's electrical contacts flawlessly align with the motherboard's corresponding pins. This connection delivers power and allows communication with other components.

Surrounding the CPU socket, you'll encounter the chipset, a collection of integrated circuits that act as a bridge between the CPU and other motherboard components. The chipset manages data flow between the CPU, memory, storage, and expansion slots, ensuring everything works in harmony. It's critical to ensure chipset compatibility with the CPU you plan to use. Modern chipsets usually incorporate the system's southbridge and northbridge functionality. Further exploration reveals the expansion slots, allowing the addition of extra hardware like graphics cards, sound cards, or networking adapters. These slots utilize various standards, most commonly PCIe (Peripheral Component Interconnect Express) slots, identified by their size and number of lanes (x1, x4, x8, x16). Faster PCIe standards will improve the data transfer rates of expansion cards, hence faster performance. Understanding this is particularly important for choosing high-performance expansion cards.

The rear panel, often overlooked, provides external connectivity. A variety of ports reside here, including USB ports for peripheral devices, audio jacks, network ports for internet connectivity, and possibly video output ports (depending on whether your board has an integrated graphics processor). The arrangement and types of ports available are largely determined by the motherboard's design. Pay careful attention to the number and type of ports you may need for your setup, as these are not easily changed.

Finally, consider the subtle details: the mounting holes for the case, the power connectors for the motherboard and CPU, and the numerous capacitors and other components that regulate voltage and ensure stable operation. Each of these seemingly insignificant elements plays a crucial role in the motherboard's overall functionality. It's vital to learn their purposes and how these components work together to create a cohesive system.

Observe the various heatsinks present on the motherboard, crucial for cooling key components. These passive coolers draw heat away from components like the chipset and voltage regulators, preventing overheating. More robust cooling solutions, potentially using active cooling (fans), may be necessary for systems with high-performance components or under heavy load.

In summary, the motherboard's anatomy is complex, yet comprehensible with systematic study. Each component, from the CPU socket to the expansion slots, plays a vital role in the computer's overall operation. Understanding the motherboard's layout and component interaction empowers you to make informed decisions about system configuration, upgrades, and troubleshooting. A deep understanding of this intricate network is a necessary step towards truly mastering your computer system.

1.2 Key Components

The motherboard's chipset acts as the central traffic controller, managing communication between the CPU and other components. Understanding its role is crucial; it's not merely a passive connector, but an active component coordinating data flow. Think of it as the sophisticated communication network within a bustling city, ensuring seamless operation. The chipset's design directly impacts the system's overall performance and capabilities, influencing factors like the number of supported USB ports, the type of RAM it can handle, and even the speed at which data transfers. A high-end chipset will offer better performance and more features, whereas a more basic chipset provides a foundation for simpler systems, usually at a lower cost.

Different chipsets cater to varying needs. For instance, a chipset designed for gaming systems might prioritize high-speed data transfer for graphics cards, while a chipset for a workstation would emphasize supporting multiple storage devices and high-bandwidth RAM. Choosing a motherboard hinges on selecting a chipset that aligns with your system's intended use. Intel and AMD are the major players in chipset manufacturing; they each offer a range of chipsets to suit different budgets and performance requirements. They frequently update their product lines; therefore, researching the latest models is always advisable before making a purchase. This research helps ensure compatibility with your CPU and other planned components.

The CPU socket, a critical component physically integrated with the chipset, determines which type of CPU the motherboard is compatible with. It's a specialized connector designed to accommodate the CPU's physical dimensions and electrical connections. Different CPU manufacturers use different socket types, meaning a motherboard with an Intel socket will not work with an AMD processor, and vice-versa. The socket's configuration also impacts the overall performance of the system. A socket offering high bandwidth ensures efficient data transfer between the CPU and other system components, optimizing the system's response times. Similarly, the number of pins in the socket signifies the number of connections and influences the processor's capabilities. Choosing the right socket is thus paramount to building a functional system.

Expansion slots provide pathways to add functionality beyond the basic motherboard configuration. The PCIe (Peripheral Component Interconnect Express) slot, for example, is an essential interface for high-performance graphics cards, sound cards, and other expansion devices. The version of PCIe used impacts the speed and bandwidth available to connected devices; higher versions allow for significantly faster data transfer rates. Multiple PCIe slots cater to various devices; the layout of these slots must be examined during motherboard selection. The motherboard layout affects the installation process and system airflow, essential elements for optimal system performance. A well-planned layout maximizes accessibility, reducing potential assembly challenges.

Consider the importance of integrated components. Many motherboards include integrated peripherals such as audio and network controllers, which eliminates the need for separate expansion cards. This simplifies system construction and can reduce costs. While integrated graphics are often sufficient for basic tasks, more demanding applications like gaming require dedicated graphics cards. It is important to consider the quality and capabilities of these integrated components as they directly influence the performance of the overall system. They provide functional baselines, but upgrading those functions often yields significant performance boosts. For many users, such integrated components are sufficient for everyday tasks.

Finally, the BIOS (Basic Input/Output System) or UEFI (Unified Extensible Firmware Interface), residing on a chip on the motherboard, is the fundamental software that governs the system's startup sequence. The BIOS/UEFI firmware loads before the operating system, initializing hardware and allowing interaction during system boot. Configurations within the BIOS/UEFI allow for adjustments, such as boot order, CPU frequencies, and fan speeds. Understanding this system is crucial for troubleshooting, configuration, and system optimization. Access to this system is usually through a specific key press during the system startup process. It offers customization capabilities allowing for better system control and tailoring settings for different user needs and system performance goals. Familiarity with BIOS/UEFI setup is an essential skill for any computer user.

2 The Central Processing Unit (CPU)

The CPU, or Central Processing Unit, is the brain of your computer. Think of it as the conductor of an orchestra, directing the flow of information and coordinating all the other components. It's responsible for executing instructions, performing calculations, and managing data. Understanding its architecture is key to understanding the entire system's performance.

This intricate architecture comprises billions of transistors, tiny switches that control the flow of electricity. These transistors are organized into functional units, such as the Arithmetic Logic Unit (ALU), which handles calculations, and the Control Unit (CU), which manages the execution of instructions. The CU fetches instructions from memory, decodes them, and then sends the appropriate signals to the ALU and other components to carry out the instructions. It's a complex dance of electrical signals, a ballet of bits and bytes. The more transistors, generally, the more powerful the CPU.

Clock speed, measured in Hertz (Hz), represents the number of cycles the CPU completes per second. Each cycle involves fetching, decoding, and executing an instruction. A higher clock speed generally translates to faster processing, allowing the CPU to handle more tasks quickly. However, clock speed isn't the only factor determining performance. The CPU's architecture, the number of cores, and cache size all play significant roles. A CPU with a higher clock speed but fewer cores might be outperformed by a CPU with a lower clock speed but more cores when handling multi-threaded applications.

Cache memory is a crucial component within the CPU architecture. It's a small, fast memory that stores frequently accessed data. Because accessing data from the cache is significantly faster than accessing it from the main RAM, using the cache dramatically improves performance. Different levels of cache exist, with L1 cache being the fastest and smallest, followed by L2 and L3 cache. Understanding cache hierarchy is important because efficient use of cache can significantly reduce the time it takes to retrieve necessary data, leading to faster overall system performance. Think of it as a chef's mise en place, having all the necessary ingredients readily available for quick access.

Overclocking involves increasing the CPU's clock speed beyond its manufacturer's specifications. This can lead to performance improvements, but it also comes with risks. Increased clock speeds generate more heat, potentially damaging the CPU if inadequate cooling is provided. Furthermore, overclocking can lead to system instability, resulting in crashes or data corruption. Before attempting to overclock, it's crucial to understand the risks and to have proper cooling solutions in place. Beginners should generally avoid overclocking, as it requires a solid understanding of system stability and monitoring techniques. Improper overclocking can irreparably damage your hardware.

The CPU socket is the physical interface on the motherboard where the CPU is installed. Different CPUs use different sockets, making them incompatible with motherboards that don't have the correct socket type. Choosing a motherboard compatible with your chosen CPU is crucial. Before purchasing any components, carefully check the specifications to ensure compatibility. Failing to do so will prevent your computer from booting correctly.

The integrated heat spreader (IHS) is a metal cover on top of the CPU die. It acts as a heat sink, drawing heat away from the CPU and dispersing it to the cooling solution, such as a heat sink or a liquid cooler. Effective cooling is essential for preventing overheating, which can lead to performance throttling or even CPU damage. Poor cooling will lead to thermal throttling, where the CPU slows down to prevent damage from overheating. Therefore, selecting an appropriate cooling solution based on the CPU's Thermal Design Power (TDP) is a critical aspect of system building.

Different manufacturers produce CPUs with varying architectures, core counts, and clock speeds. Intel and AMD are the two dominant players in the CPU market, each with its own line of processors catering to different needs and budgets. Factors like core count, clock speed, cache size, and TDP significantly influence the CPU's performance and price. Researching and comparing various models from different manufacturers is essential for selecting the CPU that best meets your needs and budget. Consider your workload—gaming, video editing, or general use—to guide your choice.

Finally, remember that the CPU is only one component in a complex system. Its performance is influenced by other components like RAM, storage, and the motherboard itself. A high-end CPU paired with slow RAM or a slow storage device might not reach its full potential. Therefore, a balanced approach to system building, considering all components, is essential for optimal performance. Optimizing other parts of your system can be just as important as upgrading your CPU. The interplay of all components contributes to the overall functionality and performance.

2.1 CPU Architecture

The CPU, the central processing unit, is arguably the most crucial component within a computer system. Its architecture dictates processing power and efficiency, influencing everything from simple calculations to complex simulations. Understanding this architecture is key to grasping a computer's capabilities and limitations. Think of it as the brain of the operation, constantly receiving, interpreting, and executing instructions. Its intricate design directly impacts the overall performance of your system. A well-designed architecture allows for smoother multitasking and faster processing speeds.

At the heart of the CPU lies the Arithmetic Logic Unit (ALU), responsible for performing arithmetic (addition, subtraction, multiplication, division) and logical operations (AND, OR, NOT). These fundamental operations underpin all computations the computer performs. The ALU receives data from registers, small, high-speed storage locations within the CPU itself. These registers act as temporary holding areas for data being processed, allowing for incredibly fast access times. The data flow between these components is incredibly rapid, crucial for maintaining processing speed. The more efficient and faster this interaction, the better the performance of the CPU.

Another key component is the Control Unit (CU). This unit acts as the CPU's director, fetching instructions from memory, decoding them, and then directing the ALU and other components to execute those instructions. The CU is like an orchestra conductor, coordinating the various parts of the CPU to work together seamlessly. Its function involves coordinating access to registers, memory, and other internal CPU components. A sophisticated CU can optimize the execution of instructions for optimal performance, especially noticeable in complex, multi-threaded applications.

Modern CPUs employ a concept called pipelining, allowing multiple instructions to be processed concurrently. Imagine an assembly line: each instruction moves through different stages of processing, and while one instruction is executing in one stage, another can enter the pipeline. This improves overall throughput and significantly increases processing speed. However, hazards like data dependencies or branch instructions can disrupt this flow, potentially reducing efficiency if not handled properly. Understanding these potential bottlenecks is crucial for optimizing code and choosing the right CPU architecture for a specific task.

Furthermore, the CPU architecture is heavily influenced by the number of cores and threads it possesses. Multi-core processors allow for parallel processing, greatly enhancing performance, especially in multitasking scenarios. Hyperthreading, a technology where each core simulates multiple threads, allows for better utilization of resources, but doesn't necessarily equal a direct doubling of performance. The effectiveness of multi-core and hyperthreading technologies depends significantly on the application's ability to utilize multiple threads efficiently.

Cache memory plays a critical role in CPU architecture. It's a small, incredibly fast memory located within or close to the CPU, acting as a buffer between the CPU and main memory (RAM). Because accessing RAM is relatively slow compared to accessing cache, frequently used data is stored in cache for quick retrieval. There are typically several levels of cache memory (L1, L2, L3), each with increasing size and slower access speeds. Effective cache management significantly impacts overall performance, often making the difference between a smooth user experience and noticeable lag.

Instruction sets, another crucial aspect of CPU architecture, determine the types of instructions the CPU understands. Different CPU architectures (like x86-64, ARM) have different instruction sets, impacting compatibility and performance. The complexity and efficiency of the instruction set influence both processing speed and power consumption. Choosing the right architecture is often dependent on the specific operating system and applications to be run. Some instruction sets are optimized for specific tasks, like scientific computation or mobile devices.

Finally, the design and integration of the CPU within the motherboard also impact performance. The CPU socket, the physical interface between the CPU and motherboard, determines compatibility. Factors such as the number of pins, the power delivery system, and the heat sink design all contribute to the overall system performance and reliability. A properly chosen and correctly installed CPU is vital for a stable and high-performing system. Neglecting these details can lead to overheating, instability, and ultimately, system failure. Therefore, paying close attention to the interplay between CPU architecture and motherboard design is absolutely essential for building a robust and efficient computer system.

2.2 Clock Speeds & Performance

The CPU's clock speed, measured in gigahertz (GHz), directly influences its processing power. A higher clock speed generally translates to faster instruction execution, resulting in quicker application loading, smoother multitasking, and enhanced responsiveness. However, it's not the sole determinant of performance; architecture, cache size, and number of cores also play significant roles. Think of it like this: a faster car (higher clock speed) will generally reach its destination quicker, but a larger engine (more cores), better transmission (cache), and aerodynamic design (architecture) all contribute to overall speed and efficiency.

Understanding clock speed is crucial for choosing a CPU. A higher clock speed isn't always better; it often comes at a cost of increased power consumption and heat generation. Balancing these factors is key to finding the optimal CPU for your needs and budget. For example, a high-clock-speed CPU might be ideal for gaming or video editing, demanding applications needing immediate processing, while a lower-clock-speed, multi-core CPU might be more suitable for tasks like compiling code or running simulations, where parallel processing is advantageous, even though it may take longer to complete each task. The type of work intended to be done on the computer greatly affects the specifications required, such as clock speed.

The relationship between clock speed and performance isn't linear. Doubling the clock speed doesn't necessarily double the performance. Architectural improvements, such as enhanced instruction sets or improved pipelining, often contribute more significantly to performance gains than simply increasing clock speed alone. Imagine two cars with identical engines; one may have a better gear ratio, allowing for faster acceleration and better overall performance, despite having the same engine speed. Similarly, architectural design is more impactful for improving the processor's overall speed.

Modern CPUs employ various techniques to optimize performance beyond simply increasing clock speed. Techniques like hyperthreading, which allows a single core to handle multiple threads concurrently, and simultaneous multithreading (SMT), which improves overall throughput, can significantly enhance processing power without a corresponding increase in clock speed. These advanced features further blur the relationship between clock speed and raw performance, emphasizing the importance of considering the whole CPU architecture, rather than solely focusing on the clock speed. In short, clock speed should be considered alongside other critical aspects of CPU architecture.

Benchmarking tools provide objective measurements of CPU performance. These tools run standardized tests to assess various aspects of CPU performance, providing scores that can be compared across different CPUs. However, real-world performance can vary depending on the specific applications being used, the other components in the system and even the software versions being used, with the influence of operating systems and other drivers. While benchmarks offer a valuable point of reference, it is more reliable to focus on the individual applications performance in a specific context. Benchmarks can be a useful guide, not a universal rule.

It's vital to understand the limitations of clock speed. Exceeding the manufacturer's specified clock speed through overclocking can lead to instability, system crashes, or even hardware damage. While overclocking can provide a performance boost, it requires careful monitoring of temperature and voltage to prevent irreversible issues. The benefits of overclocking must be carefully weighed against the risks, understanding that a higher clock speed is only one element of a comprehensive performance consideration.

Selecting a CPU involves considering many factors beyond clock speed, including the number of cores, cache size, integrated graphics capabilities, power consumption, and cost. Balancing these factors according to your specific needs is crucial for getting the best performance for your budget. For instance, a gaming system may prioritize a CPU with a high clock speed and good single-core performance, while a server might benefit more from a CPU with many cores and high multi-core performance. Prioritize the needs of your device over only focusing on one element of the hardware components.

In summary, while clock speed is an important factor affecting CPU performance, it's not the only one. Consider the entire CPU architecture, including core count, cache size, and architectural design. Pay attention to benchmark results, but remember to evaluate them in the context of your intended use. Understand the implications and risks of overclocking. By taking a holistic view of CPU capabilities, you can make a more informed decision when selecting a processor that meets your specific needs and budget. Always look at what the processor will actually be doing, rather than just focusing on singular numbers.

2.3 Overclocking Considerations

Overclocking, the practice of pushing a CPU beyond its manufacturer-specified clock speed, offers the potential for significant performance gains. However, it's a complex procedure demanding careful consideration and a thorough understanding of the risks involved. Improper overclocking can lead to system instability, component damage, and even complete hardware failure. Therefore, a cautious and informed approach is crucial.

The allure of increased processing power is undeniable, particularly for gamers and content creators. Overclocking allows you to squeeze extra frames per second from your games or render videos faster, enhancing the overall user experience. But this performance boost comes with a price: increased heat generation. Higher clock speeds necessitate increased power consumption, directly translating to more heat. This heat needs to be efficiently dissipated to prevent thermal throttling, a protective mechanism that reduces clock speed to prevent overheating, negating the benefits of overclocking. Adequate cooling is paramount – consider investing in a high-quality CPU cooler, perhaps even a liquid cooling system, especially for substantial overclocking attempts.

Before embarking on an overclocking journey, meticulous planning is essential. Research your specific CPU model and its overclocking capabilities. Manufacturer websites and online forums often provide valuable insights into the potential overclocking headroom of a particular CPU. Understanding your CPU's thermal design power (TDP) is crucial; this rating indicates the maximum power consumption under normal operating conditions. Exceeding this limit without sufficient cooling can lead to irreversible damage. Start with incremental adjustments, carefully monitoring system stability after each change. Utilize stress-testing software, such as Prime95 or AIDA64, to push your system to its limits and identify potential instability issues.

Monitoring your system's temperatures is absolutely vital throughout the overclocking process. Use monitoring software to track CPU and motherboard temperatures, ensuring they stay within safe operating ranges. High temperatures can severely shorten the lifespan of your components, making thermal monitoring an essential safety precaution. If temperatures exceed safe limits, immediately reduce the overclock or improve cooling. Never compromise the integrity of your system for a slight performance gain.

BIOS/UEFI settings play a pivotal role in the overclocking process. Familiarize yourself with your motherboard's BIOS/UEFI interface, understanding the options for adjusting clock speeds, voltages, and other relevant parameters. Begin by adjusting the CPU multiplier and core voltage in small increments. This allows for a gradual increase in clock speed while maintaining stability. Don't rush the process; patience and methodical adjustments are key. Remember, each motherboard and CPU combination will react differently to overclocking, so a generalized approach may not always be effective. Experimentation and meticulous record-keeping are important aspects of successful overclocking.

The success of an overclock depends heavily on your motherboard's capabilities. Not all motherboards are created equal in their ability to support high overclocks. Motherboards with robust voltage regulation modules (VRMs) and efficient power delivery systems generally offer better overclocking potential. Check your motherboard's specifications and online reviews to assess its suitability for overclocking. Poor VRMs can result in instability or even damage to your CPU during intensive overclocking. This is one area where investing in a high-quality motherboard can pay dividends.

Finally, remember that overclocking voids most manufacturer warranties. If something goes wrong, you might be responsible for repair or replacement costs. Weigh the potential benefits against the associated risks before attempting to overclock. Understanding the implications of overclocking is crucial, allowing you to make an informed decision. The rewards can be substantial, but the risks must never be underestimated. Responsible overclocking requires careful planning, precise execution, and a deep understanding of your system's capabilities and limitations.

3 Memory Modules (RAM)

Random Access Memory, or RAM, is crucial for your computer's performance. Think of it as your computer's short-term memory; it holds the data and instructions the CPU needs to access quickly. Unlike long-term storage like a hard drive, RAM is volatile, meaning data is lost when the power is off. This is why saving your work is so important.

Understanding RAM types is essential for choosing the right modules for your system. DDR4 and DDR5 are currently prevalent, with DDR5 offering faster speeds and lower power consumption. The numbers reflect the generation, with higher numbers indicating newer technology and better performance. Each generation offers different speeds, measured in MHz (megahertz), signifying the frequency at which data is transferred. Faster speeds generally equate to improved application responsiveness. Selecting compatible RAM is vital; using incompatible RAM can lead to system instability or even failure.

Memory capacity directly affects multitasking capabilities. The amount of RAM you have dictates how many programs you can run simultaneously without experiencing significant slowdown. More RAM generally translates to a smoother, more efficient computing experience, especially when working with resource-intensive applications such as video editing software or 3D modeling programs. Modern systems often benefit from 8GB or more of RAM for optimal performance. The actual amount needed depends heavily on your use case; a basic user may find 8GB sufficient, but a gamer or video editor will likely require much more, perhaps 16GB or even 32GB.

Dual-channel and beyond are configurations that leverage multiple RAM modules to boost performance. Dual-channel setups use two memory modules, working in tandem for improved data transfer rates. Some motherboards support even more channels, like triple or quad-channel, leading to even greater performance gains. This is achieved by utilizing multiple memory controllers within the CPU or chipset to access the RAM simultaneously. Utilizing the maximum number of channels your motherboard and CPU support is highly recommended, provided your budget allows it.

Error correction in RAM is a vital factor in system stability. ECC (Error-Correcting Code) RAM incorporates error-checking and correction mechanisms that detect and fix minor data errors. While this typically adds to the cost, it is invaluable for critical systems requiring high reliability, such as servers or workstations handling sensitive data. ECC RAM isn't always necessary for home users, but its stability benefits are undeniable for those working with substantial amounts of data or requiring high stability.

Upgrading your RAM is often a relatively simple process. The first step is identifying your motherboard's RAM type and maximum capacity. Once you know this, you can find compatible RAM modules. The upgrade process involves turning off the computer, opening the computer case, and carefully installing the new RAM modules into the appropriate slots. Your motherboard's manual will provide explicit guidance on correct placement and installation. If you are uncomfortable doing this yourself, consider seeking help from a qualified technician. Always ensure the computer is completely powered down before starting this process to avoid potentially damaging your system. Carefully follow manufacturer instructions for best results. Improper installation can lead to various problems including system boot failure.

3.1 RAM Types & Speeds

Understanding RAM speed is crucial for overall system performance. Faster RAM allows your computer to access data more quickly, leading to smoother multitasking, faster application loading times, and improved gaming experiences. Choosing the right RAM speed involves considering the motherboard's capabilities and the CPU's compatibility. For example, a high-speed RAM module won't perform optimally if the motherboard's memory controller is limited.

DDR SDRAM, or Double Data Rate Synchronous Dynamic Random Access Memory, is a common type. It transfers data on both the rising and falling edges of the clock signal, doubling the effective data transfer rate compared to earlier SDRAM. DDR2, DDR3, DDR4, and DDR5 represent successive generations, each offering improvements in speed, capacity, and power efficiency. Each generation uses different physical dimensions and voltages; therefore, compatibility is vital. Using the wrong type can damage your system.

DDR5 is the latest mainstream generation, boasting higher clock speeds and improved power efficiency compared to its predecessors. These improvements stem from several architectural changes, including a new data bus and improved signal integrity technologies. However, DDR5 modules often command a higher price and require a compatible motherboard. Consider your budget and future upgrade plans when choosing between generations. Don't rush into the newest technology if it's not cost-effective for your needs.

Understanding RAM speeds requires familiarity with specifications. Speeds are generally expressed in MHz (megahertz), representing millions of cycles per second. A higher MHz number signifies faster data transfer rates. However, the actual data transfer rate is often double the stated frequency, accounting for the double data rate capability. For instance, a RAM module rated at 3200MHz has an effective data transfer rate of 6400 MT/s (megatransfers per second).

Beyond the base speed, other factors influence overall performance. CAS latency, often expressed as CL, refers to the delay between the memory controller issuing a request and the RAM delivering the data. Lower CAS latency indicates faster access times. A RAM module with a speed of 3600MHz and a CL of 16 might be slower than a 3200MHz module with a CL of 14, especially in tasks requiring frequent small data access.

Selecting the right RAM requires careful consideration of the motherboard specifications. Check your motherboard's manual or the manufacturer's website to confirm the supported RAM types and speeds. Forcing incompatible RAM can lead to instability or even damage. Look for information on the maximum supported RAM speed and the number of RAM slots. Knowing this will allow you to determine the optimal configuration for your system. Many motherboards have detailed information available online.

Remember, RAM speed isn't the sole determinant of system performance. Other factors, including CPU speed, storage type, and overall system architecture, play crucial roles. While faster RAM can significantly improve responsiveness and performance in many applications, it's vital to strike a balance with other hardware choices and budget constraints. The most expensive isn't always the best; ensure it's actually beneficial for your usage.

Overclocking RAM can provide a performance boost, but it carries risks. Overclocking involves pushing the RAM beyond its specified speed, potentially leading to instability, data corruption, or even hardware damage if not done carefully. Many motherboards offer overclocking options within the BIOS or UEFI settings. Use caution and gradually increase the speed while closely monitoring the system's stability. Overclocking without proper understanding can cost you significant time and, if you're unlucky, even money.

Finally, upgrading your RAM is often a straightforward process. Simply open your computer case, locate the RAM slots, and carefully install the new modules, ensuring they are properly seated. Consult your motherboard's manual for specific instructions on the installation process. Always turn off your computer and disconnect the power supply before working inside the computer case. Ensure your system remains properly grounded to avoid accidental static discharge damage. Upgrading RAM is generally cost-effective and offers a noticeable performance increase.

3.2 Memory Capacity & Usage

Understanding your system's RAM capacity is crucial; it directly impacts performance. More RAM allows for more applications to run simultaneously without slowing down, leading to a smoother, more responsive user experience. Insufficient RAM forces your system to utilize slower storage devices like your hard drive for virtual memory, significantly impacting speed. Think of it like this: imagine trying to cook a complex meal with only a few small pots and pans versus having a fully stocked kitchen – the latter allows for much greater efficiency and speed.

Selecting the correct amount of RAM hinges on your intended use. Basic web browsing and word processing may only require 4GB, but video editing, gaming, and running virtual machines need substantially more, potentially 16GB or even 32GB. Consider future needs too; upgrading RAM later can be more expensive than purchasing sufficient RAM initially. Don't undervalue the importance of planning ahead to prevent future bottlenecks.

Unlike other components, RAM is relatively easy to upgrade in most systems. Simply check your motherboard's specifications to determine the maximum RAM capacity and compatible types (DDR3, DDR4, DDR5). This information, often found in your motherboard's manual or on the manufacturer's website, is non-negotiable. Then choose modules with speeds that match or closely align with your existing RAM, as this maximizes performance. There are usually specific slots on the motherboard dedicated for the RAM modules. This is where you'll be physically installing the components.

Now, let's delve into the concept of dual-channel and multi-channel memory configurations. These arrangements involve using two or more RAM modules in paired slots on the motherboard. This dramatically increases memory bandwidth, leading to substantial performance improvements, especially in tasks that heavily rely on memory transfer speeds, such as gaming or video encoding. It is similar to having two water pipes instead of one; the increased capacity allows for significantly more efficient water flow, mirroring the faster data transmission of a dual-channel setup. However, some motherboards only have one channel so if you are looking for a multi-channel setup, make sure you take the limitations into consideration.

The memory capacity isn't just about the amount; it also includes the type of RAM you are using. You should consider using the fastest RAM that your motherboard supports, as this will lead to optimal performance. But you should check the quality and reliability of the RAM module; this is an important factor to consider, and choosing reputable brands is recommended to avoid compatibility issues or failures. Don't forget to always check the specifications and make sure they are compatible, as having the wrong RAM modules can prevent your computer from working.

Properly managing your RAM usage involves understanding the processes running on your system and their memory consumption. Task Manager in Windows or Activity Monitor in macOS provide insights into which applications are using the most memory. High memory usage might indicate resource-intensive applications or potential software conflicts. Closing unnecessary applications frees up RAM, enhancing system responsiveness. Regu-

larly monitoring this ensures smooth operation and prevents slowdowns. Remember, prevention is always better than cure.
Another critical aspect is the error correction capabilities of your RAM. Error Correcting Code (ECC) memory is designed to detect and correct errors, essential for servers and other systems requiring high data integrity. While not always necessary for home PCs, ECC RAM offers increased stability and reliability, reducing the risk of data corruption. The improved reliability and stability can result in higher performance and prevent unexpected shutdowns or errors. However, ECC RAM usually is significantly more expensive than non-ECC RAM.
Finally, consider the long-term implications. Your system's memory capacity influences its longevity and upgradeability. Investing in sufficient RAM initially reduces the need for costly upgrades later. Choosing expandable motherboards and RAM allows for future capacity increases as your needs evolve. With technology constantly advancing, selecting upgradeable components ensures that your system remains efficient and relevant for longer. Always invest your money strategically and make sure that you have looked at all the options available, and remember that planning ahead helps in the long run.

3.3 Dual-Channel & Beyond

Utilizing dual-channel memory significantly boosts system performance. This configuration involves employing two identical RAM modules, working in tandem to double the bandwidth. Think of it like having two lanes on a highway instead of one; data flows much more efficiently. This leads to noticeable improvements in tasks demanding high memory throughput, such as gaming, video editing, and 3D rendering. The difference is substantial, translating to smoother, faster application response times.

However, the benefits extend beyond mere doubling of speed. Dual-channel architecture leverages optimized data transfer pathways within the memory controller, optimizing memory access times. This efficiency reduces latency – the delay between requesting data and receiving it – leading to more responsive applications and a smoother user experience. You'll find noticeably faster loading times for games and applications, an immediate improvement most users appreciate.

Moving beyond dual-channel, we encounter triple-channel and even quad-channel memory configurations. These configurations, typically used in high-end server or workstation environments, employ three or four memory channels respectively. While offering even greater performance benefits, these configurations are less common in consumer-grade motherboards. The cost increase often outweighs the marginal performance gains for the average user.

The selection of memory modules is crucial for achieving these performance boosts. The modules must be identical in terms of capacity, speed rating (measured in MHz), and timings (CAS latency and other related parameters). Using mismatched modules can severely hamper performance, potentially negating the advantages of dual-channel operation. You'll see error messages or system instability if you don't pay close attention. It's essential to carefully consult your motherboard's specifications to ensure compatibility with your chosen RAM.

Furthermore, the motherboard itself plays a critical role. The memory controller, integrated within the CPU or chipset, determines the number of channels supported. Not all motherboards support dual-channel, and some older motherboards may have limitations on the total memory capacity and speed supported. Therefore, checking your motherboard's specifications before purchasing RAM is paramount to avoid compatibility issues. Ignoring this could mean you're paying for RAM your system can't use properly.

Beyond the simple dual-channel setup, memory configurations can become increasingly complex. Flexible memory layouts allow for different combinations of RAM modules, but achieving optimal performance requires careful planning. Consider factors like the motherboard's memory slots and the maximum memory capacity it can handle. Using memory modules that exceed the motherboard's limitations may lead to system instability. A little research beforehand helps save headaches down the line.

Remember, while increasing the amount of RAM generally improves system performance, it does not directly correlate with dual-channel or multi-channel operation. The primary benefit of these multi-channel configurations lies in increased memory bandwidth, not total capacity. Therefore, while more RAM is often beneficial, optimizing the memory channels remains crucial for achieving maximum system responsiveness. Don't just focus on gigabytes; concentrate on the architecture too.

Finally, consider the impact of other system components. A slow CPU or a congested storage drive can negatively impact the overall system performance, regardless of the memory configuration. Optimizing the whole system is key. Therefore, the gains from dual-channel or multi-channel memory are maximized when all other components are working efficiently. Investing in a balanced system is always a wise move.

3.4 Error Correction

RAM, the computer's short-term memory, is crucial for speed and efficiency. However, like any system dealing with vast amounts of data, errors can creep in. These errors, even seemingly insignificant ones, can lead to system crashes, data corruption, and frustrating performance issues. Understanding and implementing error correction techniques is paramount for maintaining system stability and data integrity, particularly in systems where reliability is critical. Think of it like a meticulous librarian, constantly double-checking for misplaced books or damaged pages in a vast library.

Data corruption in RAM can manifest in subtle ways, initially leading to minor glitches or unexpected program behavior. Over time, without proper error correction, these minor issues can escalate into catastrophic failures, resulting in complete system instability. These errors can stem from various sources, such as manufacturing defects, electromagnetic interference, or even cosmic rays. Addressing these requires a proactive strategy.

One of the most prevalent error correction methods is ECC (Error Correcting Code) memory. ECC RAM utilizes additional parity bits alongside the data bits. These parity bits act as checksums, allowing the system to detect and correct single-bit errors. Think of it as a redundant backup system, providing a safety net for your data. If a single bit flips during operation, the system can instantly identify and rectify the error without impacting performance or data integrity. It adds a layer of complexity and cost but provides exceptional reliability. Servers and high-performance computing environments frequently rely on ECC RAM due to its robustness.

However, ECC RAM isn't a universal solution, and its use depends heavily on the context. It typically comes at a higher price point than non-ECC RAM, making it less appealing for everyday consumers building home computers. The added expense may not be justified for the average user, particularly if the system isn't dealing with critical applications or large datasets where data integrity is paramount. Therefore, the choice between ECC and non-ECC RAM necessitates a cost-benefit analysis.

Beyond ECC, several other strategies contribute to enhanced data reliability. Regular system maintenance, including updates and driver checks, can indirectly reduce the likelihood of RAM errors. Keeping the system clean and free of dust minimizes the risk of overheating, a common culprit for hardware malfunction. Additionally, regularly backing up important data provides another layer of protection, mitigating the potential damage caused by data corruption. Preventing errors is as crucial as correcting them.

Furthermore, understanding your system's limitations and utilizing appropriate tools for diagnostics is key. Most operating systems provide built-in memory diagnostic tools that allow users to scan for errors and address them. These tools can identify potential issues and alert you before they lead to system failure. Such proactive monitoring is crucial for identifying and rectifying potential problems before they lead to more significant repercussions. These tools are readily available and straightforward to use.

In summary, while ECC RAM offers the strongest error correction, the choice ultimately depends on the specific needs and budget. For everyday users, implementing regular maintenance routines and employing built-in diagnostic tools forms a robust strategy to minimize the impact of potential RAM errors. Remember, preventing errors is as crucial as correcting them, and a combination of proactive measures and reactive solutions offers the most comprehensive protection for your system's data. Prioritizing data integrity safeguards your work, projects, and personal files from unexpected loss or corruption, ensuring a smooth and reliable computing experience. Choosing the right approach, therefore, depends on a careful consideration of these factors and understanding the specific demands of your computational needs. Don't forget that this understanding is critical for a comprehensive approach to RAM management.

3.5 Upgrading RAM

Before beginning any RAM upgrade, meticulously consult your motherboard's manual. This invaluable resource outlines compatible RAM types, maximum capacity, and any specific installation procedures. Neglecting this step can lead to incompatibility issues, potentially damaging your system. Always verify the specifications; don't assume anything. A seemingly minor detail overlooked could render your new RAM useless.

Identifying your current RAM configuration is crucial. Open your computer's system information (usually accessible through the Control Panel or System

Settings). Note the RAM type (e.g., DDR4, DDR5), speed (e.g., 3200MHz), and capacity. This information is paramount for selecting compatible upgrades. Matching the RAM type and speed is essential for optimal performance and stability. Avoid mixing different RAM types or speeds within the same system unless your motherboard specifically supports it, and even then, proceed with caution.

Next, determine your motherboard's RAM limitations. Again, consult the manual. Motherboards have a maximum RAM capacity and may support only specific types and speeds. Exceeding these limits may result in failure. Understanding your motherboard's capabilities prevents costly mistakes and ensures your new RAM functions correctly. For instance, some motherboards are limited to dual-channel configurations; exceeding this limit won't enhance performance but could instead trigger system instability.

Choosing the right RAM upgrade demands careful consideration. Prioritize RAM with speeds matching or slightly exceeding your existing RAM, ensuring they are the same type. While faster RAM is generally desirable, ensure your CPU and motherboard support it. A faster RAM type in a system that can't handle it will simply operate at the slower speed the system can support, making the upgrade pointless. Also, consider capacity. How much RAM do you need for your applications? 16GB is generally sufficient for many users, but 32GB might be preferable for more demanding tasks like gaming or video editing. Determine your specific needs and select accordingly. Don't overspend on capacity you won't use, but equally, don't compromise on performance requirements.

The physical installation of RAM is straightforward but requires care. Ensure your computer is turned off and unplugged. Open the computer case, locate the RAM slots, and gently release the clips holding your current RAM modules. This part requires a light touch, and you will need to apply just enough force to release the clips. Once the clips are released, remove your existing RAM sticks, and place the new ones into the designated slots, making sure they are fully seated. Gently, but firmly push them down until they click into place. Remember that improper installation can cause the RAM to malfunction, and may even damage your hardware. Be meticulous, and always refer to the motherboard manual for specific instructions.

After installation, restart your computer. Check your system information again to verify that the new RAM is detected. If the system fails to boot, carefully double-check your RAM installation and refer to your motherboard's troubleshooting section. If everything is fine, the process should be smooth. If the system still does not recognize the RAM, carefully review all installation steps. There are several tools and processes available online to check your system RAM's health and integrity after the upgrade. Run them to ensure everything is working as expected. It's wise to monitor your system's performance post-upgrade and check for any reported errors.

Finally, remember that while upgrading RAM can significantly improve your computer's performance, it's not a universal solution. System bottlenecks can exist elsewhere, and only after diagnosing these would you see the full benefits of the RAM upgrade. For instance, a slow CPU or hard drive will limit the effectiveness of additional RAM. A proper assessment of the entire system is needed for optimal performance. Consider performing stress tests on your system to observe the changes and check whether or not your system is more stable with the RAM upgrade. Always back up essential data before undertaking hardware upgrades, just in case something unexpected happens.

4 Storage Solutions

Understanding the differences between Hard Disk Drives (HDDs) and Solid State Drives (SSDs) is crucial for any computer builder. HDDs, the older technology, utilize spinning platters and read/write heads to access data. This mechanical process, while cost-effective for large storage capacities, results in slower read and write speeds compared to SSDs. SSDs, on the other hand, use flash memory, allowing for significantly faster data access and a notable boost in overall system performance. The lack of moving parts also translates to improved durability and quieter operation. However, SSDs generally cost more per gigabyte than HDDs, and their lifespan, while considerable, is still finite. The choice between HDDs and SSDs often involves balancing cost, performance needs, and desired storage capacity. A hybrid approach, using an SSD for the operating system and frequently used programs and an HDD for bulk storage, offers a good compromise.

The interface technology connecting your storage device to the motherboard is just as important. SATA (Serial ATA) is a widely used interface known for its reliability and relatively high speeds, perfect for both HDDs and SSDs. However, for significantly faster performance, particularly with SSDs, NVMe

(Non-Volatile Memory Express) offers superior speeds through the PCIe bus. NVMe drives excel in applications demanding rapid data access, such as gaming and video editing. They can achieve read and write speeds multiple times faster than SATA drives. Selecting the right interface depends on your motherboard's capabilities and the type of storage drive you're using. Always check your motherboard's specifications to ensure compatibility before purchasing a new drive.

Let's delve into the specifics of SATA. SATA interfaces offer various revisions, each improving on speed and capabilities. SATA III, for example, offers theoretical speeds of 6 Gbps (gigabits per second). Understanding these revisions is essential, ensuring you're not bottlenecking your drive's potential by using an older SATA interface. It's important to choose an interface that matches the capabilities of your drive. Using a SATA III drive on a SATA II interface, for example, would limit the drive's performance. Properly configuring the SATA controller in your BIOS is essential for optimal performance. Failure to do so can result in slower than expected transfer speeds. Consult your motherboard's manual for detailed instructions on this.

Moving beyond SATA, NVMe technology significantly enhances speed and efficiency, especially relevant for SSDs. Unlike SATA which relies on a separate controller, NVMe interfaces utilize the PCIe bus, enabling significantly higher bandwidth. This translates to drastically faster boot times, application loading, and overall system responsiveness. The speed differences are often dramatic, offering considerable performance gains, particularly in demanding applications. However, NVMe drives generally come at a premium cost compared to SATA SSDs. Consider your budget and performance requirements to make an informed decision. Remember that your motherboard must support NVMe drives before you consider purchasing one.

Beyond the raw speed considerations, several factors influence storage performance. The drive's controller, a crucial component within the drive itself, plays a significant role in data transfer speeds. Some controllers are better optimized than others, influencing the overall performance you experience. Furthermore, the physical attributes of the drive itself, such as the type of NAND flash memory used in SSDs, impact speeds and longevity. Higher-end NAND technologies often offer improved performance and durability, although this comes at a higher cost. It's also important to pay attention to the drive's stated read and write speeds, which are usually presented in MB/s (megabytes per second) or IOPS (Input/Output Operations Per Second). These figures give you a good indication of the drive's performance capabilities.

Finally, consider the capacity you need for your storage devices. This decision depends on your individual needs and usage patterns. For daily tasks and general computing, a 256GB or 512GB SSD is often sufficient for the operating system and applications. For users with large media libraries or demanding storage needs, larger capacities (1TB, 2TB, or even more) are recommended. HDDs offer significantly larger capacities at lower cost per gigabyte, making them ideal for storing large archives or media collections. Consider a tiered approach, using an SSD for the operating system and frequently used programs, supplemented by an HDD for bulk storage. Careful planning in this aspect can deliver the best balance of cost, performance, and capacity. Remember to always back up your important data regularly, regardless of the type of storage device you choose. Data loss can have significant consequences, and regular backups are crucial for protecting your valuable information.

4.1 HDDs vs SSDs

Hard Disk Drives (HDDs) rely on spinning platters and magnetic read/write heads to store data. This mechanical process, while mature and relatively inexpensive, results in slower access times compared to newer technologies. Think of it like searching for a specific book in a vast library—you have to physically locate it on a shelf. This mechanical movement contributes to noise, heat generation, and a susceptibility to damage from physical shock. Data retrieval speed is significantly impacted by the rotational speed of the platters, usually measured in revolutions per minute (RPM). A higher RPM, like 7200 RPM, generally means faster access, but it also increases power consumption and noise. The lifespan of an HDD is also finite, subject to wear and tear on the mechanical components. However, HDDs still offer the advantage of high storage capacity at a lower cost per gigabyte than SSDs, making them a viable option for users needing massive amounts of archival storage. They also generally prove more resilient to data corruption than SSDs due to multiple copies of data being dispersed.

Solid State Drives (SSDs), in contrast, use flash memory chips to store data electronically. This means no moving parts, resulting in significantly faster read and write speeds, quieter operation, and greater resistance to physical shock. Imagine finding a book instantly using a digital search engine; this is akin to the speed advantage SSDs provide. This speed increase significantly impacts system boot times, application loading, and overall responsiveness. The lack of moving parts also translates to lower power consumption and heat generation. SSDs are increasingly becoming the preferred choice for operating systems and frequently accessed applications because of their responsiveness and efficiency. However, SSDs currently have a lower storage capacity per unit cost compared to HDDs, making them a less cost-effective choice for large archival storage needs. Further, the lifespan of SSDs, while typically long, is still finite; the number of write cycles for flash memory is limited, leading to potential eventual failure. And although improved, data corruption is possible, although less likely than with HDDs.

The choice between HDD and SSD ultimately depends on individual needs and priorities. For a budget-conscious user needing extensive storage space for archival data where speed isn't paramount, an HDD remains a practical option. However, for users prioritizing system responsiveness and speed in everyday tasks such as running the operating system and essential applications, the benefits of an SSD, such as faster boot times and quick application loading, significantly outweigh the higher initial cost. Consider your typical workflow: Do you frequently access large files? Is system speed crucial to your productivity? These questions will guide you toward the optimal storage solution.

The interface technology used to connect the storage drive to the motherboard also plays a role in performance. HDDs typically use SATA (Serial ATA) interfaces, while SSDs can utilize either SATA or the faster NVMe (Non-Volatile Memory Express) interface. NVMe SSDs, connected via a PCIe slot, provide considerably faster data transfer rates than SATA-based drives, making them ideal for high-performance applications. SATA, while still functional, represents a slower bottleneck in comparison. When choosing a drive, examine not only the drive itself but also the interface technology to understand potential performance limitations or enhancements. Compatibility with your motherboard's available interfaces is also crucial for functionality. Ensure that the motherboard has the necessary SATA or NVMe ports available to accommodate your selected drive. Some motherboards have a limited number of each interface, so check your motherboard's specifications to ensure you have enough slots to connect your desired drives.

In practical terms, a hybrid approach is becoming increasingly common. Many users combine both HDDs and SSDs in their systems. A relatively small, fast SSD can house the operating system and frequently used applications, while a larger, more affordable HDD provides ample space for storing data files and media. This configuration strikes a balance between speed and cost-effectiveness, allowing for both the responsiveness of an SSD and the large storage capacity of an HDD. This dual configuration maximizes the benefits of both technologies, offering an optimal storage solution for most users. Consider your budget, your application needs, and the space available within your computer case before making a decision. Weigh the pros and cons carefully, considering not just the initial purchase price but also the long-term performance implications of each technology. Ultimately, a well-informed decision will depend on a comprehensive understanding of individual requirements and available resources.

4.2 Interface Technologies

Understanding how your storage devices connect to the motherboard is crucial for optimizing system performance. This involves familiarity with the various interface technologies available, each possessing distinct characteristics impacting speed, reliability, and cost. SATA, for instance, a standard for years, offers a mature, widely compatible technology for both HDDs and SSDs, generally providing sufficient speed for most applications. However, its limitations become apparent when dealing with high-performance SSDs, especially those employing NVMe technology.

NVMe, or Non-Volatile Memory Express, utilizes the PCIe bus, offering significantly faster data transfer rates compared to SATA. This is because PCIe's parallel architecture allows for simultaneous data transmission across multiple lanes, unlike SATA's serial approach. This results in noticeably improved boot times, application loading, and overall system responsiveness. Imagine the difference between waiting several seconds for a program to open versus nearly instantaneous access. That's the power of NVMe. The higher speeds come with a price, though; NVMe drives are usually more expensive than their SATA counterparts. It's a trade-off between budget and performance you'll have to weigh carefully.

Choosing the right interface hinges on your needs. A casual user primarily focused on everyday tasks like web browsing and email might find SATA perfectly adequate, offering a good balance between cost and functionality. The differences in speed won't be hugely noticeable in their day-to-day experience. However, for a serious gamer, video editor, or someone working with large datasets, the jump to NVMe can be transformative, making a tangible improvement to workflow efficiency. The faster speeds are especially beneficial when dealing with large files or complex processes.

Beyond SATA and NVMe, other interfaces exist, though less common in consumer-grade systems. These often cater to specialized applications or older hardware. Understanding these is less critical for beginners but provides a broader context of storage interface evolution. Staying updated on the latest technologies ensures you're making informed choices when building or upgrading your system. For example, newer standards continually emerge, pushing the boundaries of storage performance.

Furthermore, consider the physical connection itself. While both SATA and NVMe use distinct physical connectors, both require proper seating and secure connections within the motherboard's designated slots. A loose connection can lead to data corruption or system instability, highlighting the importance of careful installation and regular inspection. Don't overlook the small details; proper installation is crucial for reliable performance and longevity.

Finally, remember to check your motherboard's specifications to ascertain which interfaces it supports. Not all motherboards are created equal. Some may only offer SATA ports, while others boast multiple NVMe slots for maximum flexibility. Knowing your motherboard's capabilities before purchasing storage devices avoids compatibility issues and wasted expenditure. Always verify compatibility; it is the key to trouble-free operation. This will save you time and frustration in the long run, ensuring your system operates at its peak efficiency. Always read the manuals, and you'll be set. Thorough research and planning pay off when building or upgrading a computer.

5 Graphics Processing Unit (GPU)

Understanding the GPU's role is crucial for anyone building or upgrading a computer. Unlike the CPU, which excels at general-purpose calculations, the GPU is specifically designed for parallel processing, making it ideal for handling the massive computations required for graphics rendering. This parallel architecture allows it to perform many calculations simultaneously, significantly speeding up tasks involving large datasets—rendering 3D images, video editing, and even scientific simulations.

Integrated GPUs, often found on lower-end motherboards, are directly embedded within the CPU. They provide basic graphics capabilities, sufficient for everyday tasks like web browsing and office work, but lack the power for demanding applications. Think of them as a modest paintbrush for simple drawings. Dedicated GPUs, on the other hand, are separate physical cards that plug into a PCIe slot on the motherboard, offering vastly superior performance. These are akin to an artist's complete studio, capable of creating masterpieces in any style. Their higher processing power comes at a price, of course, both in terms of cost and power consumption.

Choosing between an integrated and dedicated GPU boils down to your needs. Budget-conscious users satisfied with modest performance may find integrated graphics sufficient. However, gamers, video editors, or anyone working with graphically intensive software will require the significantly increased power of a dedicated GPU. The difference is substantial, allowing for higher resolutions, smoother frame rates, and vastly improved overall visual experience. The impact is similar to that of using professional recording equipment versus using a cellphone camera.

Several metrics help quantify GPU performance. Clock speed, measured in MHz, indicates how fast the GPU's internal components operate. Higher clock speeds generally lead to increased performance, but other factors also play a significant role. Memory bandwidth, determining how quickly data can be transferred to and from the GPU, is equally crucial. A higher bandwidth means faster data processing, resulting in improved performance. Finally, the number of CUDA cores (in NVIDIA GPUs) or Stream Processors (in AMD GPUs) directly impacts the parallel processing capabilities; more cores mean more calculations can be performed concurrently. Consider these factors when comparing GPUs, much like you would compare engines in automobiles, based on horsepower and other performance characteristics.

Selecting the right GPU depends heavily on intended use. For casual gaming at moderate resolutions, a mid-range GPU can provide excellent performance. High-end gaming and professional applications, such as video editing or 3D modeling, necessitate a top-tier GPU with superior performance capabilities.

Budget, power consumption, and physical size are all critical factors influencing the choice. Larger GPUs often deliver superior performance but demand more power and occupy more space within your computer case. It's crucial to balance your needs with these limitations.

Beyond basic specs, consider features such as ray tracing support, which enhances the realism of rendered images and videos, or DLSS (Deep Learning Super Sampling), a technique that boosts frame rates without sacrificing significant visual quality. These advanced technologies aren't essential for basic tasks, but they can transform the experience for demanding applications like gaming and virtual reality. This is like comparing a basic car with a vehicle equipped with advanced safety features and assisted driving capabilities. The added functionality often enhances the overall user experience and enhances its capabilities. Careful research and comparison are vital, helping you to choose the GPU that best fits your budget and meets your needs. Reviews and benchmarks are a helpful tool that will help you make an educated choice. Don't let the technical jargon intimidate you; focus on what's really important for your needs.

5.1 Integrated vs Dedicated GPUs

Integrated graphics processors, often abbreviated as iGPUs, are built directly onto the motherboard's chipset. This integration offers a cost-effective solution, especially for budget-conscious systems or those with less demanding graphical needs. Think of it as a basic, built-in feature, like the speakers on your laptop; they work, but they won't replace a high-quality sound system.

The performance of an iGPU is intrinsically linked to the chipset's capabilities and shares system resources with the CPU and RAM. This shared resource allocation means performance can be noticeably impacted when running graphics-intensive applications alongside other processes. You might experience lag or slowdowns when gaming or working with high-resolution video editing software. Consequently, the iGPU is ideal for basic tasks like web browsing, document editing, and light media consumption. Its limited processing power is a key differentiator and a significant factor to consider when choosing a system configuration.

Dedicated GPUs, however, operate independently. They are separate physical components that connect to the motherboard via a PCIe slot. Imagine a dedicated, powerful musician versus a built-in keyboard – entirely different levels of performance. They possess their own dedicated memory (VRAM) and processing units, allowing for significantly improved graphical performance compared to integrated solutions. This dedicated power translates to smoother gameplay, faster rendering times in design applications, and the ability to handle high-resolution displays with ease. The increased processing power comes at a cost, both financially and in terms of physical space. Dedicated GPUs are more expensive and require a larger power supply, and their size can be a limiting factor in smaller form factor systems.

The choice between integrated and dedicated GPUs hinges primarily on your intended use. If your computer needs only handle basic tasks like web browsing or word processing, an iGPU suffices. The lower cost and reduced power consumption make it an attractive option for everyday use. It offers sufficient processing power for less demanding applications. You won't find yourself waiting long for everyday tasks. However, for gaming, video editing, 3D modeling, or other graphically demanding tasks, a dedicated GPU is absolutely essential. The noticeable performance increase justifies the higher cost and power requirements. This investment allows for a smooth, efficient, and high-quality experience.

Consider the long-term implications as well. An integrated graphics solution lacks the upgradeability of a dedicated GPU. As your needs evolve, upgrading your GPU involves upgrading the entire motherboard, a process far more complex than simply replacing a dedicated graphics card. Dedicated GPUs allow for future-proofing your system. You can easily swap out an outdated GPU for a newer, more powerful model. This upgradability ensures your system remains relevant and capable of handling future demands. With changing technologies and escalating software needs, it's a significant factor when planning your system's lifecycle.

Ultimately, the decision comes down to a careful assessment of your budget and your intended use cases. If your primary need is simple computing, then iGPU is a practical, cost-effective choice. It perfectly meets the needs of a standard user. But if you anticipate using your computer for graphically intensive tasks, a dedicated GPU is the way to go. For seamless performance and future-proofing your system, it's an investment worth considering. The superior performance and upgradeability ensure a smooth experience now and in the years to come. The initial cost differential is often outweighed by the long-term benefits and enhanced capabilities. The crucial point is to align your hardware choice with your needs, avoiding unnecessary expenses and ensuring optimal performance. Careful consideration will significantly influence your overall computer experience. Choose wisely to best suit your present and future needs, striking the ideal balance between performance and practicality.

5.2 GPU Performance Metrics

Understanding a GPU's capabilities goes beyond simply knowing its brand and model. Several key metrics offer a deeper insight into its performance, crucial for informed decision-making, especially for computationally intensive tasks like gaming or professional graphics applications. These metrics provide a quantitative assessment of the GPU's processing power, memory bandwidth, and overall efficiency. Ignoring them can lead to suboptimal performance or even incompatibility with your chosen applications.

Clock speed, measured in MHz (megahertz), represents the frequency at which the GPU's core processes data. A higher clock speed generally translates to faster processing, but it's not the sole indicator of performance. Modern GPUs often employ boosting algorithms that dynamically increase clock speeds based on workload and thermal headroom. Therefore, understanding both the base clock and the boost clock is essential for a complete picture of the GPU's potential. Pay attention to benchmarks and reviews that highlight both values rather than just focusing on a single number.

Stream processors, also known as cores or shaders, are the fundamental processing units within a GPU. The number of stream processors directly correlates to the GPU's parallel processing capability. More stream processors generally allow for faster rendering of complex scenes, higher frame rates in games, and quicker completion of computationally intensive tasks. However, the architecture and efficiency of these processors are also crucial; more isn't always better. A GPU with fewer but more efficient stream processors might outperform one with a larger quantity but less efficient design.

Memory bandwidth, measured in GB/s (gigabytes per second), represents the speed at which data can be transferred between the GPU's memory and its processing units. High memory bandwidth is vital for smooth performance, especially in high-resolution games or applications that require substantial texture and data processing. A bottleneck in memory bandwidth can significantly limit overall GPU performance, even if other metrics are high. Consider the memory type (GDDR6, GDDR6X) and its bus width as well; this plays a substantial role in achieving maximum bandwidth.

Memory capacity, measured in GB (gigabytes), refers to the amount of video RAM (VRAM) available on the GPU. VRAM acts as short-term storage for textures, models, and other data required for rendering. Insufficient VRAM can result in slowdowns, stuttering, and texture pop-in, particularly at higher resolutions and detail settings. The required VRAM amount depends heavily on the resolution and the graphical fidelity of the applications being used. High-resolution gaming or professional applications requiring extensive texture detail will demand greater VRAM capacities. Therefore, consider your future needs carefully.

Thermal design power (TDP), measured in watts, represents the maximum power consumption of the GPU under normal operating conditions. A higher TDP generally indicates a more powerful GPU with a greater processing capacity, but also higher power consumption and potentially more heat generation. This affects your power supply requirements, cooling solution, and overall system efficiency. It's prudent to ensure that your power supply has sufficient wattage, and your cooling system can adequately dissipate the heat generated, to avoid throttling or system instability. Moreover, consider the efficiency rating of the GPU, as this impacts its power consumption under different loads.

Fillrate, expressed in billions of pixels per second (GPixels/s), is a metric indicating the speed at which the GPU can color pixels. It's directly related to the number of rasterization units and their clock speed. A higher fillrate results in smoother frame rates, especially in scenes with many polygons or complex lighting effects. Although fillrate is important, it's not the only factor affecting visual quality; other aspects such as texture filtering and shading techniques also play crucial roles.

Finally, benchmark results provide a holistic assessment of GPU performance across different applications and tests. Look for benchmarks using applications relevant to your needs—games for gaming, professional software for 3D modeling, and so on. Remember that benchmark scores are relative; they should be compared within a specific generation and set of testing conditions to gain accurate comparisons and expectations. Sites like 3DMark, and user reviews from communities and forums provide comprehensive benchmarks and real-world feedback that can guide your purchasing decision and align with your expectations. Always be critical in your evaluation of benchmark data and consider the source's credibility and testing methodology.

5.3 Selecting a GPU

Choosing the right GPU can significantly impact your system's performance, especially for graphically intensive tasks like gaming or video editing. Understanding your needs is paramount. Consider the resolution of your monitor; a higher resolution demands more processing power. If you're aiming for 4K gaming at high frame rates, you'll need a much more powerful GPU than for 1080p gaming at lower settings. Your budget plays a crucial role too; high-end GPUs can be expensive. Before diving into specific models, establish a clear budget to avoid exceeding your financial limits.

Next, think about the types of applications you'll primarily use. For gaming, look for GPUs known for their high frame rates and smooth performance in popular game titles. For professional applications like video editing or 3D modeling, prioritize GPUs with strong rendering capabilities and large amounts of VRAM (Video RAM). VRAM acts as the GPU's short-term memory, holding textures and other data needed for real-time processing. The more VRAM you have, the more demanding applications you can run smoothly. Insufficient VRAM can lead to performance bottlenecks and slowdowns.

Let's delve into specific performance metrics. The GPU's core clock speed indicates how fast its processing units operate; higher clock speeds generally mean better performance. However, clock speed isn't the only factor. The number of CUDA cores (Nvidia) or stream processors (AMD) also significantly impacts performance. More cores enable parallel processing, allowing the GPU to handle multiple tasks concurrently. Memory bandwidth measures the rate at which data can be transferred to and from the GPU's memory. Higher bandwidth translates to faster data access and improved overall performance, especially in games and applications that demand large amounts of data transfer.

Now, let's address the issue of integrated versus dedicated GPUs. Integrated GPUs are built directly into the CPU, offering basic graphics capabilities suitable for everyday tasks like web browsing and video playback. However, they lack the power for demanding applications. Dedicated GPUs, on the other hand, are separate cards that plug into the motherboard's PCIe slot, providing significantly higher performance for gaming, video editing, and other graphically intensive tasks. Dedicated GPUs are essential if you plan on performing any visually demanding work. Selecting a dedicated GPU is almost always a better choice for any performance related work.

Consider the size and cooling solution of the GPU. High-end GPUs can be quite large and generate considerable heat, requiring robust cooling systems. Ensure your computer case has adequate space and airflow to accommodate the chosen GPU's dimensions and prevent overheating. Overheating can lead to performance throttling or even damage to the GPU. Also, check the power requirements of the GPU. High-end GPUs often require a significant amount of power, and you may need to upgrade your power supply unit (PSU) to meet its demands. Always consult the GPU's specifications to determine its power requirements.

Finally, research reviews and benchmarks. Websites and publications dedicated to PC hardware often publish detailed reviews and benchmarks of various GPUs. These resources can provide valuable insights into a GPU's real-world performance, helping you make an informed decision based on actual testing and user experiences. Pay close attention to benchmarks that align with your intended use case (gaming, video editing, etc.). Remember that newer generations of GPUs generally offer better performance than older ones, provided you're staying within the same price range and choosing similar specifications. Understanding these metrics will enable you to make a well-informed choice. Careful consideration will guarantee that your new GPU fits seamlessly into your current system, providing a performance boost that meets your expectations.

6 Expansion & Connectivity

PCIe slots are your motherboard's high-speed express lanes. Think of them as dedicated roadways for data-hungry devices like graphics cards and high-performance SSDs. Different PCIe versions offer varying bandwidths, much like highways with different speed limits. A newer generation PCIe slot, like PCIe 4.0 or 5.0, will significantly outperform an older PCIe 2.0 slot, enabling faster data transfer rates crucial for demanding applications such as gaming or video editing. Knowing the version of your PCIe slots is essential when choosing compatible expansion cards. Incorrect pairing can lead to performance bottlenecks or outright incompatibility. Always check the specifications of both your motherboard and the expansion card before installation.

USB ports, ubiquitous on motherboards, provide connectivity for a wide range of peripherals. From keyboards and mice to external hard drives and printers, USB ports offer a standardized and versatile interface. However, not all USB ports are created equal. You'll encounter various USB standards – USB 2.0, 3.0, 3.1, 3.2, and even the newer USB4 and USB-C – each offering different data transfer speeds and power delivery capabilities. Faster USB standards allow for quicker data transfer, crucial for high-bandwidth devices like external

SSDs. Understanding these differences helps you optimize your peripheral connections and choose the right port for the job. For example, connecting a high-speed external SSD to a USB 2.0 port will severely limit its performance potential. Selecting the appropriate port is a simple but often overlooked aspect of maximizing system efficiency.

Networking interfaces, another key aspect of motherboard connectivity, provide the pathway for your computer to interact with the wider digital world. Ethernet ports offer a wired connection, typically delivering more stable and faster network speeds than Wi-Fi, making them ideal for applications requiring consistent bandwidth like online gaming or video conferencing. However, many modern motherboards also integrate Wi-Fi and Bluetooth capabilities, providing wireless connectivity for convenient device pairing and network access. When choosing between wired and wireless options, consider the trade-offs between convenience and speed. A wired connection generally prioritizes speed and stability but sacrifices mobility, while a wireless connection offers flexibility at the cost of potential speed reductions and instability.

Beyond PCIe slots and USB ports, many motherboards offer other expansion options. These could include internal headers for connecting case fans, additional USB ports, or even specialized connectors for audio systems. These options allow for customization and expandability to tailor your system to your specific needs. Refer to your motherboard's manual to fully understand the capabilities of these internal connectors and how they can enhance your system. For instance, connecting extra case fans via internal headers allows for better cooling, which is particularly useful for overclocking your CPU or

GPU. This chapter aims to empower you to utilize the full potential of these features.

Understanding BIOS/UEFI settings is crucial for managing your system's hardware and optimizing its performance. BIOS (Basic Input/Output System) or UEFI (Unified Extensible Firmware Interface) is the firmware that initializes your computer's hardware before the operating system boots. These settings allow for configuring various hardware aspects like boot order, storage device configurations, CPU overclocking parameters, and power management options. Navigating the BIOS/UEFI interface can seem daunting, but it is a valuable tool for fine-tuning your computer's behavior. Accessing the BIOS/UEFI settings usually requires pressing a specific key (often Delete, F2, F10, or F12) during the boot process. Consult your motherboard's manual for the correct key combination. While many settings are best left untouched unless you understand their implications, configuring boot order, for example, allows you to select the primary boot device. The BIOS/UEFI menu is a powerful tool that allows you to control and tweak many of your motherboard's settings. Take time to explore the options carefully, always exercising caution, and refer to the motherboard manual or online resources if unsure about any particular setting. Mastering BIOS/UEFI settings is a crucial step in fully controlling and optimizing your computer system. Remember, safety always comes first. If you are unsure, seek assistance from experienced individuals.

6.1 PCIe Slots & Devices

Peripheral Component Interconnect Express (PCIe) slots are expansion slots found on motherboards, providing high-speed communication pathways for various devices. Understanding their capabilities is crucial for maximizing your system's potential. These slots utilize a serial data transmission protocol, significantly faster than older parallel technologies like PCI. This speed advantage allows for the seamless integration of high-bandwidth devices, ranging from graphics cards that power demanding games and professional applications to network interface cards (NICs) capable of handling high-speed data transfer.

The PCIe standard isn't monolithic; it's evolved over several generations. Each generation boasts an increase in bandwidth, reflected in both data transfer speeds and the overall performance enhancement they bring to connected devices. Identifying the generation of a PCIe slot is easy—look for the 'x' followed by a number (e.g., PCIe x16, PCIe x4). This number dictates the number of lanes, or pathways, available for communication. More lanes mean more bandwidth, directly impacting the performance of the connected device. A PCIe x16 slot, commonly used for high-end graphics cards, provides substantially more bandwidth than a PCIe x1 slot, often employed for sound cards or low-bandwidth expansion cards.

When choosing devices for your PCIe slots, always check for compatibility. Simply matching the physical size isn't enough. You must ensure the device supports the generation of your PCIe slot. Trying to use a modern PCIe 4.0

device in an older PCIe 3.0 slot will work, but you'll only leverage the older generation's slower speeds. The reverse is usually impossible; a PCIe 3.0 device likely won't be compatible with a PCIe 4.0 slot at all. Always consult the specifications of both your motherboard and the device you intend to install. Manufacturers often provide clear compatibility information on their websites and product documentation.

Beyond generation and lane count, pay attention to the physical slot's size. The length of PCIe cards varies widely, therefore, ensure your case has sufficient clearance to accommodate the card's dimensions. Careless installation can lead to overheating or damage to the device or the motherboard itself. Prior to installation, consult your motherboard's manual to determine the precise location and capabilities of each PCIe slot. Some motherboards may have specific slots designated for particular functionalities like high-bandwidth graphics cards; using the 'wrong' slot could limit performance.

Moreover, consider power requirements. High-performance devices such as powerful graphics cards or high-end network adapters often require supplemental power connectors directly connected to your computer's power supply. These connectors provide the necessary power for the card to operate effectively, preventing performance throttling or system instability caused by insufficient power. Always check the power requirements listed in the device's specifications; neglecting this can lead to malfunctions and damage. Furthermore, ensure adequate airflow around your PCIe cards, especially high-performance ones that generate significant heat. This can be achieved through proper case ventilation and the use of case fans or coolers specifically designed for graphics cards. Ignoring this can result in thermal throttling, significantly diminishing the device's performance and potentially shortening its lifespan. Proper installation is key to maximizing performance and preventing damage. Always power down your computer and unplug the power cord before handling any internal components. Carefully align the PCIe device with the slot and gently push it in until it firmly clicks into place. Avoid using excessive force. If it doesn't click, double-check for alignment issues and try again. After installation, secure the card to the case using the provided screws to prevent accidental dislodging during operation. Reassembling your PC requires methodical care to ensure components are secured appropriately.

PCIe slots offer immense versatility for extending the functionality of your computer system. They facilitate connectivity for a wide variety of devices, ranging from essential components to specialized add-ons that tailor the system to specific needs. Therefore, understanding PCIe's capabilities and correctly selecting and installing devices in these slots is a fundamental aspect of building a high-performance and reliable computing environment. By adhering to these guidelines, you can optimize your system for years of trouble-free operation. Remember, meticulous planning and careful execution are vital for a successful and harmonious interaction between the motherboard and its expansive PCIe ecosystem. Always consult your specific motherboard and device documentation for detailed compatibility information and installation procedures.

6.2 USB & Other Ports

Understanding your motherboard's connectivity options is crucial. The variety of ports available dictates your system's versatility and expandability. Knowing the differences between these ports allows you to choose peripherals that best suit your needs and avoid compatibility issues. This knowledge isn't just for advanced users; even basic computer literacy requires a grasp of these fundamental components.

USB ports, ubiquitous on modern motherboards, are your primary interface for connecting a huge range of peripherals. From mice and keyboards to external hard drives and printers, USB's versatility is unmatched. However, not all USB ports are created equal. You'll encounter various USB standards, including USB 2.0, USB 3.0, USB 3.1, and USB 3.2, each offering different data transfer speeds. USB 2.0, while still common, has significantly slower speeds compared to its more recent counterparts. USB 3.0 and its successors provide much faster transfer rates, crucial for large files and high-bandwidth devices. Identifying these versions – often indicated by color-coding or specific labeling on the ports themselves – is vital to optimize your peripheral performance. Remember that while backward compatibility generally exists (you can use a USB 2.0 device on a USB 3.0 port), using a newer standard significantly boosts the performance of compatible devices.

Beyond USB, other ports offer specialized connectivity. These include HDMI and DisplayPort for video output to monitors, Ethernet ports for wired network connections, and audio jacks for speakers and microphones. Understanding the functionality of these ports is essential for correct peripheral connection. Mismatching a device with its intended port could lead to malfunctions or outright failure. For example, connecting a high-resolution monitor to a DisplayPort port will typically yield better visual quality than using an older HDMI port, depending on the monitor and graphics card capabilities. A common mistake is to assume all HDMI ports are the same; however, different versions exist, which can affect maximum resolutions and refresh rates. Thoroughly checking your motherboard specifications and your peripherals' requirements before connecting them is always a good preventative measure.

Consider your future needs when evaluating your motherboard's port selection. If you plan on expanding your setup significantly, ensure your motherboard has enough ports to accommodate your peripherals. A lack of USB ports, for instance, can limit your ability to connect multiple devices simultaneously. This becomes particularly important in scenarios with a multitude of peripherals, such as a gaming setup or a home studio, requiring numerous input and output devices. Planning ahead minimizes potential bottlenecks and ensures smooth operation.

The arrangement of the ports also matters. Ports clustered closely together might impede the use of larger devices. Likewise, the physical location of the ports, whether on the front or back panel, can affect accessibility and usability. For instance, often front-panel USB ports are convenient for quick device connections, but their connection might be limited to USB 2.0 due to the wiring limitations. A thorough examination of your motherboard's specifications will reveal this detail.

Finally, don't overlook the importance of maintaining your ports. Dust and debris can accumulate, causing connectivity problems. Regular cleaning, using compressed air, will ensure optimal performance and longevity of your motherboard and connected devices. A clean motherboard also prevents overheating and ensures efficient energy usage. Preventive maintenance extends the lifespan of your system and prevents costly repairs. This simple step is often overlooked but is just as important as understanding the technical aspects of your ports. Remember that properly maintaining your system is an essential part of responsible technology ownership. A well-maintained system is a more productive and reliable system.

6.3 Networking Interfaces

Motherboards offer several ways to connect to networks, each with its own advantages and disadvantages. The most common are Ethernet and Wi-Fi. Ethernet, using the RJ-45 connector, provides a wired, stable, and generally faster connection. This is ideal for situations requiring consistent high bandwidth, like online gaming or large file transfers. However, it requires a physical cable, limiting mobility. The speed of an Ethernet connection depends on both the network card and the network infrastructure; Gigabit Ethernet (1 Gbps) is common, while newer standards like 10 Gigabit Ethernet (10 Gbps) offer significantly higher speeds, though these require specialized cabling and hardware. Consider cable quality, as poor cabling can severely impact performance. A faulty Ethernet cable can lead to intermittent disconnections or slower speeds, so always ensure you use a high-quality cable and test it regularly.

Wi-Fi, on the other hand, provides wireless connectivity, offering great convenience and mobility. It's suitable for laptops and other portable devices, allowing you to access the network from anywhere within range of your router. However, Wi-Fi connections can be affected by interference from other devices or physical obstacles, resulting in lower speeds and instability compared to Ethernet. The speed of your Wi-Fi connection depends on several factors including the Wi-Fi standard (802.11ac, 802.11ax/Wi-Fi 6/6E, etc.), the router's capabilities, and the distance between your device and the router. Newer Wi-Fi standards provide significant speed and range improvements, but compatibility needs to be considered with both your motherboard and your other devices. Wi-Fi is also susceptible to security vulnerabilities, requiring strong passwords and potentially other security measures.

Beyond Ethernet and Wi-Fi, some motherboards integrate other networking technologies. Bluetooth, for instance, enables short-range wireless communication with devices like keyboards, mice, and headphones. While not strictly a networking interface in the same sense as Ethernet or Wi-Fi (it doesn't connect to the internet directly), it simplifies connectivity for peripherals. The version of Bluetooth implemented matters, with newer versions offering improved speed and range. This is particularly useful for wireless peripherals and data synchronization. Ensure that your motherboard supports the latest Bluetooth versions for optimal performance and compatibility. Moreover, some high-end motherboards might incorporate other less common interfaces, such as Thunderbolt, which offers exceptionally high bandwidth for connecting external storage devices or displays.

Choosing the right networking interface depends on your needs and priorities. For situations requiring maximum stability and speed, wired Ethernet is the preferred option. If mobility and convenience are paramount, then Wi-Fi is the more practical choice. It is also beneficial to understand the different standards for each type to select the appropriate hardware. For example, purchasing a motherboard with support for the latest Wi-Fi 6E standard allows for greater speeds and reduces interference. Consider the implications of your choice carefully. If you're building a desktop PC that will primarily be used in a fixed location, Ethernet should be given greater consideration. If you need a highly portable system with a considerable amount of wireless communication, Wi-Fi 6E offers superior performance. Remember always to check your motherboard's specifications to ensure compatibility with the desired networking technologies. Careful consideration of each networking interface will ensure that your system meets your connectivity needs.

Finally, properly configuring your network interface is crucial for optimal performance. This involves installing the correct drivers for your network card, selecting the appropriate network settings in the operating system, and ensuring your network security is up to par. Outdated or incorrect drivers can lead to instability, slow speeds, or connectivity issues. Always download the latest drivers from the manufacturer's website. Your operating system's network settings allow you to configure various aspects of your network connection, such as IP addresses, DNS servers, and network protocols. Properly configuring these settings ensures that your computer can communicate effectively with other devices on the network. Furthermore, strong passwords and firewalls are essential for protecting your network from unauthorized access and malicious attacks. Neglecting these important aspects can significantly impact the security and performance of your system. Therefore, proper configuration and maintenance of your networking interfaces is essential for a smooth and secure computing experience.

6.4 Expansion Options

Motherboards offer a variety of expansion options, significantly impacting system functionality and future upgrade potential. Understanding these options is crucial for building a versatile and adaptable computer. The most prominent expansion slots are the PCIe (Peripheral Component Interconnect Express) slots, categorized by their physical size and data transfer speeds. PCIe x16 slots, for example, are designed for high-bandwidth devices like graphics cards, while PCIe x1 and x4 slots accommodate lower-bandwidth peripherals such as sound cards or network interface cards. Choosing the right slot for a specific device is critical; using a high-bandwidth device in a lower-bandwidth slot severely limits its performance.

Selecting expansion cards involves more than just fitting them into the right slot. Each card has specific requirements and must be compatible not only with the motherboard's PCIe standard but also with the system's power supply. High-performance cards, such as high-end graphics cards, often require additional power connectors directly from the power supply. Failing to provide adequate power can lead to system instability or even hardware damage.

Therefore, always check the card's specifications and your power supply's capabilities before installation. The installation process itself generally involves securely seating the card into the appropriate slot and fastening it down with a retaining clip. Remember to refer to both your motherboard and expansion card manuals for precise instructions to avoid damaging the components.

Beyond PCIe slots, motherboards include various other expansion options. These include additional USB ports, either integrated onto the motherboard itself or via expansion cards. USB standards vary in data transfer speeds, power delivery capabilities, and connector types. Understanding these differences is crucial for choosing the correct ports for different peripherals. For example, high-speed data transfer requires using a USB 3.0 or later port, while power-hungry devices may require a USB port with a higher power delivery capacity. The location of these ports also matters; some may be conveniently located on the back panel while others might be internally accessible.

Furthermore, other expansion options might include M.2 slots for NVMe solid-state drives, providing extremely fast storage solutions. These slots offer a compact and high-speed alternative to traditional SATA interfaces. However, ensure your motherboard supports the specific M.2 standard and drive size before purchasing. Improper compatibility can lead to incompatibility issues, rendering the drive unusable. Another important element is the onboard audio. While most motherboards include integrated audio, dedicated sound cards offer significantly improved audio quality for users demanding superior sound reproduction, and this upgrade might require the use of a PCIe slot.

Remember, expansion options significantly impact system performance and adaptability. Before purchasing any components, consult your motherboard's specifications to ensure compatibility and confirm the availability of the necessary slots and connectors. Understanding the technical specifications and limitations of each expansion option is paramount to building a well-rounded and future-proof system. Choosing the right expansion cards is as important as selecting the core components, because they expand the functionality and increase the lifespan of your computer, allowing it to be adaptable to future technological advancements. Always prioritize compatibility to prevent potential issues. Carefully read the documentation provided with your motherboard and expansion cards to ensure optimal performance and longevity.

6.5 BIOS/UEFI Settings

Accessing your motherboard's BIOS or UEFI (Unified Extensible Firmware Interface) is crucial for several reasons; it's your gateway to configuring fundamental system settings, and many critical adjustments are impossible to make otherwise. This includes controlling boot order, which dictates the device from which your system starts, altering the CPU's operating parameters and managing the system's power consumption. Understanding your BIOS/UEFI is fundamental to efficiently utilizing your system's hardware.

The process of entering the BIOS/UEFI varies slightly across motherboard manufacturers, but generally involves pressing a specific key (often Delete, F2, F10, F12, or Esc) repeatedly during the initial power-on self-test (POST). The exact key is usually displayed briefly on the screen during POST. If you miss this, restarting your computer and trying again is always an option; don't panic if it takes a couple of attempts. Your motherboard's manual offers the most reliable information on this procedure.

Once inside, you'll be greeted by a text-based interface, often with menus and submenus organized by category. Navigation typically involves using the arrow keys to move between options, the Enter key to select, and possibly other keys for saving changes and exiting. Every BIOS/UEFI will present a unique layout and some will be more visually appealing than others. Familiarise yourself with the navigation and don't worry; it's surprisingly intuitive once you get the hang of it.

One of the most vital settings you'll encounter is the boot order. This determines the sequence in which the system attempts to boot from various devices, such as hard drives, SSDs, USB drives, and optical drives. Correct boot order configuration is essential, especially when installing operating systems or troubleshooting boot problems. Incorrect settings here can prevent your computer from starting at all. Carefully select your preferred boot device from the provided options to ensure you boot from the correct drive.

Another crucial setting is the CPU configuration. Here, depending on your motherboard and processor, you might find options to adjust the CPU's core voltage, multiplier, and clock speed. However, exercising extreme caution is paramount; incorrect adjustments can damage your components. Overclocking, while potentially enhancing performance, introduces risks of instability and overheating. Unless you're very experienced, it's best to leave these settings at their default values to avoid serious issues.

Beyond CPU settings, you'll likely find options related to memory management, such as XMP (Extreme Memory Profile) settings. This profile automatically configures your RAM to run at its rated speed, maximizing its performance. Enabling XMP, when appropriate, is usually beneficial and usually involves selecting your RAM kit's profile from a provided list. Note that this may require careful consideration if you have multiple RAM modules with varying profiles.

Furthermore, power management settings within the BIOS/UEFI allow adjustments to the system's power consumption and operational states. For example, you can configure the system to enter sleep or hibernation modes automatically after a specified period of inactivity. These settings can influence power efficiency and battery life, particularly on laptops. Explore these options and see how the settings can better suit your personal needs and preferences.

Security settings are also present in most BIOS/UEFI interfaces. Features like Secure Boot and TPM (Trusted Platform Module) help secure your system against malware and unauthorized access. Whether you enable these features might depend on your security requirements and operating system; you might want to check your operating system's security preferences before altering these BIOS/UEFI settings.

Finally, remember to save your changes before exiting the BIOS/UEFI. Usually, there's a specific option to save and exit; failing to do so means any modifications won't be applied. Saving your settings is essential to make the changes permanent and ensure the next boot uses the settings you adjusted.

Always consult your motherboard's manual for detailed instructions and specific settings available within your BIOS/UEFI. It's a valuable resource that provides insights into the nuances of your motherboard's capabilities. The manual will clarify any uncertainty and provide guidance for safe configuration. Don't hesitate to look for it online if you don't have a physical copy. This chapter provides a general overview; specific options and their functionalities may vary widely between different motherboard models.

7 Troubleshooting & Maintenance

A failing system often manifests subtly; a slow boot, intermittent freezes, or unexpected shutdowns. These aren't always catastrophic, but ignoring them can lead to data loss or complete system failure. Understanding the root cause is crucial. Start by identifying the specific symptoms. Is the computer unresponsive? Does it repeatedly restart? Are certain applications crashing? Precise symptom identification is your first step toward effective troubleshooting. Power supply issues are common culprits. An insufficient or failing power supply can starve the motherboard and other components, leading to unpredictable behavior. Check the power cord connections, both at the wall outlet and the computer's power supply unit (PSU). Visually inspect the PSU for any signs of damage, such as bulging capacitors or burnt components. Consider using a power supply tester to measure voltage output. If you suspect a power supply problem, it's safer to replace the unit rather than risk damaging other components.

Overheating is another significant concern. Excessive heat can damage components, leading to instability or failure. Ensure proper airflow within your computer case; clean out dust build-up using compressed air, paying close attention to fans and heat sinks. Monitor CPU and GPU temperatures using monitoring software. If temperatures are consistently high, consider applying new thermal paste to the CPU and GPU heat sinks. Remember, always disconnect the power supply before working inside the computer case.

BIOS/UEFI issues can cause a range of problems, from boot failures to hardware incompatibility. Access the BIOS/UEFI settings by pressing the appropriate key during startup (usually Delete, F2, F10, or F12, depending on the motherboard manufacturer). Check the boot order to ensure your primary storage device (hard drive or SSD) is listed first. Also, verify that the date and time are correctly set, as incorrect settings can sometimes lead to unexpected errors. If you encounter problems within the BIOS/UEFI, consider resetting it to default settings.

Memory problems, often manifested through system instability or random crashes, can be tricky to diagnose. Running a memory test, such as MemTest86, is recommended. This free utility thoroughly checks your RAM for errors. If the test reveals issues, replace the faulty RAM module(s). Be sure to follow the motherboard manufacturer's instructions for proper installation. Improperly seated RAM is a frequent cause of system instability. Always check to ensure the RAM clips are securely fastened.

Hardware conflicts can arise when multiple devices contend for the same resources. This could involve peripherals or expansion cards. Carefully examine the Device Manager in Windows or System Information in macOS to identify any conflicts or errors. If conflicts are detected, try disabling or uninstalling the conflicting devices. Updating device drivers can also help resolve many conflicts. Regularly check for driver updates from the manufacturer's website. This proactive approach often prevents potential issues.

While hardware problems are frequent, software issues can also trigger system instability. Malware or corrupted system files can interfere with the motherboard's operation. Regularly update your operating system and antivirus software. Consider running a system file checker to identify and repair corrupted files. A clean operating system install, while a more drastic measure, may resolve persistent software-related issues. Remember to back up your important data before undertaking such a procedure.

Regular maintenance is crucial for the long-term health of your system. Cleaning dust from the inside of your computer case is essential. Accumulated dust restricts airflow, leading to overheating. Regularly update drivers and software for improved stability. In addition, periodically check the connections of all components, ensuring that they are secure. Loose connections can cause intermittent problems, ranging from display issues to network connectivity failures. Proactive maintenance prevents many potential issues and extends the lifespan of your system. This approach is far more efficient than emergency troubleshooting.

Finally, understanding your motherboard's specifications is critical. Consult your motherboard's manual for detailed information about its capabilities, supported components, and troubleshooting tips. The manual often contains crucial information not easily found elsewhere, including detailed pinouts, jumper settings, and troubleshooting guides. Keep this valuable resource readily available. Referencing it before attempting any advanced troubleshooting procedures can prevent accidental damage to your system. Proactive consultation often avoids costly mistakes.

7.1 Common Motherboard Issues

Motherboard malfunctions can stem from various sources, often presenting as seemingly unrelated problems. A seemingly simple boot failure might actually indicate a failing capacitor, a subtle power supply issue, or even a corrupted BIOS. Identifying the root cause requires a systematic approach, starting with the most obvious symptoms. Don't jump to conclusions; thorough investigation is crucial for effective troubleshooting. Remember, patience is key—rushing the process can lead to further damage.

Power problems frequently manifest as complete system failure or erratic behavior. A failing power supply unit (PSU) might not provide enough power for the motherboard and its components, leading to unexpected shutdowns or instability. Insufficient voltage can cause the system to crash, display error messages, or simply refuse to boot. Check your PSU's power output using a multimeter, ensuring it meets your system's requirements. Look for bulging or leaking capacitors on the motherboard itself—these are strong indicators of a power-related problem. If you discover such defects, professional repair is usually necessary, as replacing individual capacitors requires specialized knowledge and tools.

Overheating is another major culprit behind motherboard failures. Components like the chipset and the voltage regulator modules (VRMs) generate significant heat, particularly under heavy load. Inadequate cooling can lead to thermal throttling, instability, or even permanent damage. Ensure your case has proper airflow, with sufficient fans to exhaust hot air and keep components within safe operating temperatures. Monitor your system's temperatures using software like HWMonitor. If temperatures consistently exceed manufacturer's recommendations, consider adding more fans, replacing your CPU cooler, or applying new thermal paste to improve heat transfer. Remember, even a seemingly minor temperature increase can accumulate damage over time.

Sometimes the issue isn't about a malfunctioning component, but rather poor contact. Loose connections can disrupt signals and cause all sorts of issues, from intermittent boot failures to unexpected shutdowns. Carefully inspect all connections, ensuring that every cable is firmly seated and that no pins are bent or damaged. Pay close attention to the CPU and RAM connections—even a slightly misaligned module can result in a system failure. Consider reseating all components, including the CPU, RAM, and expansion cards. A visual inspection with a magnifying glass might help to pinpoint subtle issues.

BIOS corruption can lead to a complete system failure, preventing even the initial boot sequence. This can happen due to power surges, faulty firmware updates, or hardware failures. A corrupted BIOS usually results in a blank screen or a series of cryptic error messages. In some cases, a BIOS update might resolve the problem, but always proceed with caution and follow manufacturer instructions precisely. If a BIOS update fails, professional repair or motherboard replacement might be necessary.

Damaged components, such as short-circuited traces or damaged capacitors, can render the motherboard unusable. This can result from liquid spills, physical impacts, or even electrostatic discharge (ESD). Visual inspection is essential here. Look for any visible signs of damage, such as burnt marks or discolored areas. If you identify significant damage, the board may be beyond repair. Unfortunately, motherboard repair, especially at the component level, requires specialized skills and equipment; in many cases, replacement is more practical and cost-effective.

Finally, remember to take preventative measures. Ground yourself before working on any computer components to prevent ESD damage. Maintain a clean and organized workspace to avoid accidental damage. Regularly clean out your computer case to remove dust buildup, which can impede cooling and lead to overheating. Consider investing in a surge protector to protect your system from power fluctuations. Regular software updates for your BIOS and drivers can improve stability and reduce the risk of unexpected problems. A well-maintained system is much less likely to experience motherboard issues. Proactive care is far simpler than reactive repair.

7.2 Diagnostic Techniques

Visual inspection is your first line of defense. Carefully examine the motherboard for any obvious physical damage: bent pins on the CPU socket, bulging capacitors, burn marks, or broken traces. These are often readily apparent and can pinpoint the source of the problem. Don't forget to check for any loose components; a simple reseating might solve the issue. A magnifying glass can be incredibly helpful in identifying subtle damage, especially on densely populated boards. Remember, safety first; always disconnect the power supply before handling any internal components.

Next, consider the power supply. A faulty power supply is a common culprit behind seemingly unrelated motherboard problems. Ensure your PSU delivers the correct voltage and amperage, and check its connections to the motherboard. A multimeter can accurately measure voltage output, allowing for a precise assessment of its functionality. If you suspect a power supply issue, it's often best to test with a known-good power supply unit before replacing the motherboard.

Testing the motherboard components requires a systematic approach. Start by checking the CPU. Make sure it's properly seated and the cooler is adequately installed. Then, test the RAM modules. Reseat them; this simple act often resolves seemingly complex issues. Use the system's memory diagnostic tools and consider using a memory testing program for a thorough assessment of RAM integrity. Don't forget to check other peripherals – hard drives, graphics cards, and expansion cards. These components can cause issues if improperly seated or faulty.

If visual inspection and basic checks fail to reveal the problem, proceed to more advanced diagnostics. A POST (Power-On Self-Test) card can be invaluable. This small device provides a numeric code that can often pinpoint the source of a system failure during boot, providing specific information on the components that aren't functioning correctly. Consult the manufacturer's documentation or online resources to decipher the code.

Utilizing a motherboard diagnostic jumper is another effective method. These jumpers typically have different positions, each activating a different diagnostic mode. Check your motherboard's manual to identify the correct jumper configuration for the troubleshooting steps. This will guide you towards a pinpointing of the fault. Remember, misplacing the jumper could damage your motherboard, so exercise care and consult the manual thoroughly.

Beyond hardware diagnostics, software tools play a significant role. Your operating system's built-in diagnostic utilities provide insights into system health, identifying potential issues with hardware, drivers, or software conflicts. Utilize system event logs to pinpoint errors and use resource monitors to investigate high resource consumption from processes or hardware issues. Many operating systems include sophisticated error checking and logging that can pinpoint sources of instability or failure, even if only after the fact. Remember to update all relevant drivers to the latest versions. Outdated or corrupted drivers can often cause unforeseen problems.

Advanced diagnostic tools, like specialized motherboard testers, can offer highly detailed analysis. These tools directly assess the motherboard's circuitry, checking for shorts, open circuits, and other internal issues. They are more expensive and often require technical expertise to operate. Such tools provide data that go beyond the capabilities of built-in system diagnostics, giving a thorough overview of the motherboard's status, but require an additional investment. However, in cases where other troubleshooting methods are unsuccessful, their use can be invaluable.

Finally, don't underestimate the value of seeking professional help. While many motherboard issues are diagnosable and fixable at home, some problems require specialized tools and expertise. If you've exhausted all other options and still can't pinpoint the problem, it might be cost-effective and time-saving to consult a computer repair technician. They have extensive experience and access to specialized equipment, allowing them to efficiently diagnose and repair complex motherboard issues quickly. The cost of their services might be less than purchasing several replacement parts that may not resolve the issue.

7.3 Preventative Measures

Regular cleaning is paramount. Dust accumulation acts as an insulator, trapping heat and potentially leading to overheating, which can damage components. Use compressed air to gently blow away dust from all areas of the motherboard, paying close attention to the fan blades and heatsinks. Avoid using excessive force, which could damage delicate parts. A soft brush can help remove stubborn dust particles. Remember, preventative maintenance is far less expensive than repairs.

Static electricity poses a significant threat. Before working on your computer, always ground yourself by touching a grounded metal object. Static discharge can instantly fry sensitive components. An anti-static wrist strap provides extra protection, connecting you to the ground and preventing potentially damaging electrical discharges. These are inexpensive and readily available. Don't underestimate the destructive power of even a minor static shock; it can ruin a perfectly good motherboard.

Proper power management is crucial. Power surges can cause immediate and irreversible damage to your motherboard. Invest in a surge protector, a device that absorbs excess voltage and prevents it from reaching your computer. Unplug your computer during thunderstorms. These precautions are small investments compared to the cost of replacing a damaged motherboard. Moreover, habitually shutting down your computer completely instead of using sleep or hibernate modes reduces wear and tear on the components.

Maintaining a stable operating temperature is critical for longevity. Ensure adequate airflow around your computer case, using case fans if necessary. Clean the case fans regularly. Overheating is a major cause of motherboard failure, and proper cooling reduces this risk significantly. Monitor your CPU and GPU temperatures using monitoring software; if they frequently exceed recommended levels, investigate solutions such as adding more case fans or better heatsinks.

Regular software updates are essential. BIOS updates often include bug fixes and performance improvements that can enhance the stability and reliability of your motherboard. Check your motherboard manufacturer's website periodically for updates specific to your model. However, proceed cautiously; incorrect BIOS flashing can render your motherboard unusable. Carefully follow the manufacturer's instructions. Similarly, updating your operating system regularly patches vulnerabilities that could lead to malware infections.

Avoid excessive stress on components. Never forcefully insert or remove components. Handle all components with care, paying attention to delicate connections. Gently push the components into their slots, checking that they are correctly installed. Similarly, avoid rough handling of your computer; drops or impacts can cause damage to the motherboard and other internal components.

Understanding your motherboard's specifications is key. Consult your motherboard's manual to understand its power requirements, operating temperature ranges, and supported components. Avoid exceeding these limits. For instance, installing a high-wattage CPU without sufficient cooling may lead to overheating and damage. Operating within the specified parameters extends the life of your motherboard considerably.

Protecting against liquid damage is vital. Never consume food or beverages near your computer. Spills can cause short circuits and irreparable damage. If a spill occurs, immediately unplug your computer and seek professional assistance. Liquid damage is often difficult to repair, and preventative measures are far more effective than attempting a post-incident recovery. Moreover, ensure that your computer is placed in a dry and well-ventilated area, away from potential sources of spills or humidity.

Regular back-ups are not directly related to motherboard care, but they are crucial for data preservation. If a catastrophic motherboard failure does occur, you won't lose your precious data. Implement a reliable backup strategy, using external hard drives, cloud storage, or a combination of methods. Regular backups ensure that even in the event of hardware failure, your valuable data is safe and recoverable. It's peace of mind, and a practical preventative measure against data loss. Losing your work due to hardware failure is far more devastating than replacing the motherboard itself.

Finally, always remember that a proactive approach is far more effective and less costly than reactive repairs. Regular preventative maintenance significantly reduces the likelihood of encountering costly motherboard issues. By following these simple steps, you'll significantly increase the lifespan and reliability of your motherboard. Your investment will last longer, and you'll avoid the headaches and expenses associated with unforeseen hardware failures. Remember, prevention is always better than cure, and in the world of computer hardware, this adage holds particularly true.

www.ingramcontent.com/pod-product-compliance
Lightning Source LLC
LaVergne TN
LVHW022015200926
848481LV00007B/2403
9798349286230